colourful **hanging baskets**

& *other containers*

colourful **hanging baskets**

& *other containers*

Tessa Evelegh
& Debbie Patterson

Quadrille

page 1: The copper tones of a wire basket complements the sepia-coloured leaves of a trailing pelargonium.

page 2: The delicate dancing appeal of hellebores brings lightness to the darkest winter days. Plant a pair in rusting urns to accentuate their natural elegance.

page 3: Planted in a container, the distinctive beauty of Dicentra spectabilis can be admired at close quarters.

opposite: Ground-cover roses such as this tiny-leaved Rosa 'Nozomi' can be grown in containers as standards to produce a cloud of blossom-like blooms.

Publishing Director: Anne Furniss
Art Director: Mary Evans
Project Editor: Jackie Matthews
Design Assistant: Ian Muggeridge

First published in 1997 by
Quadrille Publishing Limited
27-31 Charing Cross Road
London WC2H 0LS

This paperback edition first published in 1999

Cataloguing in Publication Data: a catalogue record for this book is available from the British Library.

ISBN 1 899988 14 9

Printed in Hong Kong

contents

introduction

The joy of hanging baskets and containers is that you can put colour exactly where you want. For a fuller splash of colour, pack plants in their containers then position them around the garden to create highlights. Use window boxes to brighten sills, hang baskets either side of the front door to provide a welcome, or decorate the patio with potted shrubs. The easiest way to create impact is to plan coordinated schemes, often relying on bedding favourites. In addition, unusual and stunning visual arrangements can be easily achieved by ingeniously marrying different types of containers with interesting plant combinations. To keep the colour going right through the year, you can replant containers with fresh shades for each new season as soon as plants have finished flowering. The eye-catching, colour-themed planting suggestions illustrated in this book can be either copied as they are or used as inspiration for your own experiments with plant combinations and unlikely containers. Each chapter is devoted to a different colour, offering plenty of planting ideas that span all the seasons. The chapter at the beginning of the book gives practical information on compost and planting as well as discussing ways of using a variety of containers or improving their appearance. A section at the end of the book provides the names of plants that can be used in place of many of the plants featured should these be unavailable in your area or if you prefer to experiment with alternative planting ideas.

This well-planned grouping ensures a sophisticated combination of black and tan from springtime right through the summer. When it is first planted, the Euphorbia griffithii 'Fireglow' provides the tan, and as this fades, field poppies and violas take over.

tools & equipment

Baskets and containers come in all shapes, sizes and materials. Collect together an assortment of attractive and interesting items so that you have plenty to choose from to form the basis of a new planting idea.

The sheer beauty of a hanging basket or window box overflowing with colour brings pleasure in itself. And that pleasure is all the greater in the knowledge that this delightful show is a reward not only for creative planting but for careful nurturing. Container gardening is not difficult, but the artificial conditions of the plant holder, with limited space for compost and the dense planting, mean that feeding and watering cannot be left to nature in quite the same way as in the rest of the garden. Yet, given the right conditions plus plenty of tender loving care, hanging baskets and other containers will reward with a luscious show of seasonal colour. To ensure you get the most from your baskets and containers, this chapter contains advice on composts and planting as well as general maintenance and aftercare. Different and unusual options for baskets and containers which contribute so much to a planting design, are discussed as well as ways of embellishing and decorating them.

containers unlimited

Garden centres sell all sorts of containers: baskets that hang from chains or ropes, wall-mounted baskets often shaped like mangers, window boxes, troughs and pots in endless designs made from many different materials including metal, wood, terracotta and stone. In addition, almost any container designed for another purpose can be used for planting as long as it is equipped with drainage holes. If you want to use an improvised container, bear in mind that the smaller the volume of compost it can hold, the more likely it is to dry out during hot weather. Aim to choose containers that are at least 23cm (9in) wide.

Hanging baskets

The most basic hanging baskets are made from plastic-coated wire, but there is also a wide choice of more substantial cast-metal designs that range from simple suspended baskets to wall-mounted baskets and mangers in various shapes and sizes. These are all constructed so that plants can be planted through the sides, as well as at the top, to create a lush floral cascade which completely covers the basket by the end of the summer. More expensive, though ultimately much more beautiful, are the decorative wirework baskets that were popular in Victorian times and which

Almost any type of container can be used for planting an interesting arrangement. Those with solid bases will need several small drainage holes drilled through the bottom with the appropriate tool for the material. Make sure the container is stable before you start drilling.

are still made today. These are particularly effective when planted to last through the seasons as they look wonderful even when their contents have yet to attain their best or are even past it.

Plenty of other types of open-structured container can be adapted to make stunning hanging planters. Among these are decorative wire baskets, old colanders, garden sieves, even willow shopping baskets, all of which have integral holes, and so are particularly suited to the purpose. Almost anything that can be hung will also do, though you will need to make drainage holes in the bottom if it is solid.

Just about any open-structured container can be used as a hanging basket. Wicker shopping baskets (far right) are ideal, but even a discarded birdcage can be adapted. This one (right) is crammed with Felicia amelloides *variegated and* Convolvulus tricolor *'Blue Ensign'.*

Fitted with ropes or chains around their rims, terracotta, glazed ceramic and plastic pots also make excellent hanging containers. When the plants are fully grown, the effect differs from that of a conventional hanging basket as there is no opportunity for side planting and also because the container is likely to be deeper. The result is that, rather than being obscured by a lush covering of blooms and foliage, the container becomes an integral part of the whole arrangement, offering a greater sense of structure.

Suspending baskets

Baskets are usually suspended from ropes or chains and hooks. Chains are often sold with hooks already fitted for attaching to the basket. Alternatively, you can improvise with ropes or fancy chains. When hung from a wall, baskets need sturdy brackets that are at least as long as the radius of the basket to hold them clear of the wall. Fixings need to be strong and secure.

Liners

Hanging baskets that do not have solid sides must be lined to retain the compost. You can make your own liner from a sheet of plastic, which will also reduce loss of water from the basket, but you will need to cut drainage slits in the bottom. Alternatively, for standard-shaped baskets, you can buy a purpose-made liner.

Keep a stock of chains and hooks in the tool shed in readiness for improvising a suspended basket.

There is a wide variety of these, often made to fit particular sizes of basket. Some, such as those made from compressed fibre, are pre-moulded in the shape of a basket, and if you want to plant through the sides, you will need to make holes using a tool such as an apple corer. Other liners, like foam-plastic or coconut fibre types, are disc-shaped with slits from the centre to the sides so they can take up the shape of the basket while allowing for side planting. The flaps often overlap a little, but the plants can be tucked between them.

Some purpose-made hanging baskets have a built-in water reservoir underneath the main body of the basket. The water is transferred to the compost by way of a capillary mat which is placed over the reservoir at planting time. This certainly helps to prevent the compost from drying out during very hot weather, but it would not provide a sufficient reserve of water to last more than a weekend.

Standing containers

All sorts of containers can be used to form the basis of any number of imaginative plantings: old beer barrels, galvanized baths, discarded sinks, abandoned birdcages, wooden boxes or fruit crates, holed buckets and even old chimney pots can all be pressed into service and given a new lease of life as long as you are able to make drainage holes in the bottom.

Liners, available in a number of different materials, are essential for holding the compost in baskets.

decorating containers

Coloured glass baubles laid on the surface of the compost provide an unusual decorative mulch for red-tinged Imperata cylindrica *'Rubra'.*

Even the simplest plantings can be turned into something special by decorating the container to create a unique finished arrangement. It is easy to do using paint, shells, old crocks, beads, raffia and string or even chickenwire, to list just a few options.

Baskets

Purpose-made hanging baskets usually have wire or metal struts, and are designed to be eventually covered by the plants growing in them. When they are freshly planted, however, the sides are there for all to see, and it is at this time that you may like to add your own decorative touches. You could weave raffia in and out of the struts for a woven effect, for example, or wire on sparkling chandelier drops, pierced shells, glass beads or even tiny bells, fixing them in position with wire or garden string so they hang from the rim.

Solid-sided containers

Pots and solid-sided containers provide more scope for decoration than hanging baskets, and as they will not necessarily become hidden by the plants they contain there is arguably a greater need for embellishment. Paint provides endless possibilities and allows you to create wonderful effects very easily, even if you do not feel you are a natural artist. Simply painting pots in plain colour so that they coordinate with the plants that are to be grown in them or with their surroundings can add impact to the whole garden. For a bolder statement, a large half barrel can be transformed into a sophisticated receptacle for a large plant. The handsome tub shown opposite (top left) is easy to

reproduce with the help of some masking tape and stencils (see page 85). For a simpler result, you could stick with the purely geometric and simply paint broad stripes on your pots. Paint them all over in one colour, then, when the paint has dried, mark off stripes using masking tape and repaint the alternating stripes in a contrasting colour. When you remove the masking tape, you will be left with perfect stripes. Paint can also be used to create *faux* effects, and you can buy kits to create verdigris and antique bronze finishes.

If painting is not your forté, you may be happier adding decoration in the form of stuck on objects such as shells, mosaic or beads. Shells are particularly suited for use outdoors. Designed to protect sea creatures from the roughest seas, they can cope easily with the rather tamer elements of the garden. Ecologically, the best shells to use are those from the fishmonger, and there is a wide choice available, ranging from mussels and oysters to scallops. Shells are easily fixed in position using a glue gun and hot wax sticks.

Old crockery, broken up into small pieces, can be used to mosaic pots and containers. Cut the china into fragments using tile snips, then work out a design and fix it in position with exterior tile adhesive. When the adhesive is dry, finish with tile grout to fill the gaps between the fragments.

An incredibly simple yet effective way of adding decoration to a container is to make up a necklace of beads or pebbles using string or wire and drape it around the top of a pot. Or gather fine-mesh chickenwire around a pot, fix it in position with a band of garden wire, then add a decorative tie of raffia.

Ways of adding distinction and individuality to containers and plantings are seemingly endless. Here are just a few ideas, ranging from the smartly painted heraldic approach of the half barrel (top left) to the seashells simply stuck on an ordinary rectangular container (below right). Ruby chard is bound up with a hank of raffia (below left) while blue bottles upturned on canes add an interesting dimension to bright red pelargoniums (geraniums) (top right).

the ingredients

Hanging baskets and containers look best when they are closely planted, and this, combined with the limited space, means that a lot is expected of the growing medium. As well as a good compost, plants in containers need more fertilizer and more water than those growing in the ground, and all these things need to be considered when planting up.

Compost

The basic growing medium needs to be bulky enough to support the roots, be water retentive yet free draining, and it needs to contain enough fertilizer to give the plants a good start. Also, it should be sterile, so that weeds, rot and disease are not inadvertently transferred to the container or basket and subsequently affect the planting. Ordinary garden soil would not be able to fulfil all these requirements and, unless you want to mix up your own planting mixture, it is best to buy a proprietary compost specially formulated for hanging baskets and containers.

There are two basic types of compost: soil-based ones and soilless ones, which are peat- or coir-based. The soil-based composts, such as the John Innes formulae, are better able to hold nutrients, but as they are heavier they are less suitable for hanging baskets, especially larger ones. Choose John Innes No 3 for vigorous summer growth. The soilless composts are cleaner and lighter, but they hold both nutrients and water less efficiently, and if they are allowed to dry out, they can be difficult to re-wet. In this case, the best solution is to immerse the whole container in water until the compost is thoroughly moist. In addition, bulb fibre, a peat-based mixture, may be used for bulbs.

As many manufacturers produce hanging basket and container composts, there is plenty of choice. These are more likely to be well supplied with nutrients than many other kinds of compost, and some also include moisture-retaining agents. Check the packaging for details of exactly what the compost contains. For lime-hating plants, you will need to use ericaceous compost.

Moisture-retaining granules

When planting hanging baskets and containers, it is a good idea to mix moisture-retaining granules into the compost to reduce the rate of water loss over the first few days or even weeks, just when the plants are establishing themselves. These polymer crystals can hold up to 400 times their own weight in water.

Slow-release fertilizers

Slow-release fertilizer granules can also be mixed into the compost when planting up. These have a porous coating and when water passes through it, some of the fertilizer is released and dissolves in the compost. In some products, the amount of fertilizer released is controlled by the temperature of the compost, so more is released during hotter weather, when it is most needed. Different formulations are designed to work over different lengths of time, say two, three or even five months. Choose a formulation to suit the length of time the planting is expected to last. Alternatively, you can use one to give an early boost to a planting, and then add extra feed regularly once the slow-release fertilizer is exhausted.

Hand tools

The tools required for planting containers and then maintaining them are simple but essential.

A short-handled trowel is most useful, though a fork can be used for levelling the compost in newly planted containers. Choose a good quality tool.

Good secateurs are important for pruning, shaping and removing dead foliage. Invest in some that have strong handles that are comfortable to use and have strong sharp blades; do not try to economize by buying a cheap pair. Clean cuts damage the plants less and promote healthy re-growth. Secateurs are also essential for the sometimes daily task of deadheading.

Containers need copious watering, especially during hot and windy weather, when they can dry out within a few hours. This is most easily done when water is literally 'on tap'. If possible, add a rose or shower-head fitting for a gentle action. A more powerful jet of water can over-wet some areas while leaving others dry, and it can disturb the compost. Sometimes it is easier to water containers and baskets using a long-spouted watering can fitted with a rose, as it can get among the plants and water the compost directly. This is also useful for watering after planting up.

If you need to train woody plants, such as roses, this is best done using soft garden string which is likely to rot before it can do any permanent damage to the plant stems. If forgotten about, wire ties can cut into branches and stems as they grow.

Compost, moisture-retaining granules, fertilizer, good drainage and basic tools are essential for containers.

the secrets of success

Planting up containers is fairly straightforward but if you bear the following advice in mind you will increase your chances of success. And once the plants are installed and growing well, regular maintenance – watering, feeding and deadheading – will repay your efforts many times over.

Good planting

Before re-using a container, scrub it thoroughly to remove any pests, eggs, fungal spores or other disease. Always cover the drainage holes of pots and containers with crocks or large pebbles to prevent the compost from silting out and clogging up the holes.

Water plants thoroughly before planting them to prevent dehydration. As very wet compost is difficult to handle, allow the plants to drain for at least 45 minutes before removing them from their pots; some experts like to leave them to drain overnight. If the plant does not come out of its pot easily, place two fingers on either side of the stem, upturn the pot and give it a sharp tap.

There should always be a minimum 2.5cm (1in) layer of compost below the roots of any plant and, depending on the size of each rootball, the depth may well be much deeper. The rootball should be bedded into the container allowing a 2.5cm (1in) watering depth between the surface of the compost and the rim of the container. Work out how deep the compost should be by trying the plant in the container and then fill to the correct depth with compost. Push plenty of compost between and around each plant and press it down firmly with your fingertips.

An awkwardly shaped rootball may fit into the container better if it is loosened first. To avoid tearing roots to fit the plant into a tight space, plunge it into a bucket of water and gently work away some of the compost. Make sure the roots are surrounded by plenty of compost in the new container.

Try to plant showy summer hanging baskets as early as possible in the season to allow the plants time to bed in well before flowering. If you plant before the frosts are over, keep the containers in a greenhouse or conservatory at night, until late spring.

Watering

Correct watering of containers is crucial. Plants that are underwatered soon wilt and drop their lower leaves and flowers. Overwatered ones drop both young and old leaves, and their flowers become mouldy. Waterlogged compost rots the roots.

It is important to wet all the compost evenly, so water until water flows out through the bottom of the container. To avoid waterlogging, however, wait until the top of the compost is dry to the touch before watering again. On hot dry days containers may need watering twice or even three times a day. The smaller the container, the more vigilant you need to be. Water either early in the morning or in the evening as leaves can be scorched by the sun if water touches them. Try to water at compost, not leaf, level. For large plants with plenty of low leaf cover, it may be helpful to remove one or two of the lower leaves. In winter, you will probably need to water only once a week. Do not water in frosty weather.

Deadheading and pruning

While watering, look over summer baskets to check whether any plants need attention. At the beginning of the season, many plants benefit from having their growing tips pinched out to promote bushiness. Once plants are blooming, deadheading encourages a new flush of flowers. Some plants, such as petunias, are inclined to become leggy and so benefit from being cut back. Cut back one plant a week to about 10cm (4in) above the compost to avoid denuding the container. Trailing Surfinia® varieties of petunia, which produce flowers down their stems do not need to be cut back.

Feeding

Container-grown plants need regular feeding. Slow-release fertilizers come in the form of granules (see page 16), pellets or sticks, which are pushed into the compost. They usually provide fertilizer over two to three months and sometimes up to five. If you did not add a slow-release fertilizer to the compost when planting up, there are several other options. Liquid feed, a concentrate that is applied when watering, can be used weekly during the fast-growing summer months. Some types are specially formulated for hanging baskets. Seaweed extract can be used either to drench the roots or to spray the leaves. Never apply foliar feed in bright sunshine as this could damage the leaves. Powdered or granular feed is sprinkled over moist compost, pricked in, and then watered in. Apply it when the leaves are dry and not on a bright day so that the plant can be thoroughly watered to clean any stray feed off the leaves.

pretty in pink

**STRAWBERRIES
AND CREAM**
Several varieties of
strawberry, planted
with antirrhinums in
pink and cream,
deep pink to cerise
trailing verbena and
variegated ground
ivy were the
inspiration for this
enchanting summer
basket.

Ranging from the faintest blush to the richest carmine, pink is undeniably feminine, offering a softness to the garden whatever the season. Making an unassuming entrance in spring with primulas, tulips and hyacinths, pink's favourite season is summer when it explodes in all its shades on almost every type of flower. By autumn, the tones are hot with crimson asters, heathers and sedum vying with the oranges, bronzes and yellows that are more typical of the season. By winter, pink seems to have burnt itself out, and is almost absent from the garden. You can create striking baskets and containers using just one shade of pink, or you can combine several shades and textures in pink for even more impact.

strawberries & cream

You will need

3 *Fragaria* 'Serenade'

6 *Verbena* 'Delight'

6 *Glechoma* (syn. *Nepeta*)
   hederacea 'Variegata'
   (variegated ground ivy)

3 *Fragaria* 'Strawberries
   and Cream'

Antirrhinum majus, dwarf,
   pink and cream

Large wire hanging
   basket, about 50cm
   (20in) diameter

Pot or bucket to stand
   basket on while working

Sphagnum moss, to line
   basket

Plastic sheet

Scissors

Compost

Season Summer

When to plant Late spring

Site Light shade

Strawberries make enchanting plants for hanging baskets. Their leaves provide full cover, while their flower- and fruit-bearing shoots trail decoratively over the side. There are plenty of interesting varieties to choose from, and the combination of two or more of these can provide a gloriously rich planting. Here, a creamy-variegated variety, itself called 'Strawberries and Cream', offers pretty foliage, while 'Serenade', with its simultaneous flowering and fruiting, provides interest throughout the summer season. Antirrhinums in both pink and cream have been planted in the centre of the basket to add height, while trailing verbena and variegated ground ivy lend extra texture.

❶ Water the plants thoroughly and allow them to drain. Stand the basket on a pot or bucket to keep it stable. Line it with sphagnum moss, then with plastic sheet and cut planting slits around the sides of the basket for some of the 'Serenade' strawberry shoots and verbenas. Place a 7.5cm (3in) layer of compost in the bottom of the basket.

❷ Cut a square of plastic large enough to wrap around one of the side plants. Lay a plant on the square, diagonally from corner to corner, and gently roll the plastic around the plant to form a cone with the point at the flower end.

❸ Carefully thread the pointed end through one of the slits in the plastic liner, from inside the basket, and pull the cone through until only the roots are on the inside. Then gently pull the plastic away, leaving the plant in position.

❹ Plant all the side plants in the same way. Fill the basket with compost so that the tops of the rootballs of the remaining plants stand 2.5cm (1in) below the rim of the basket when placed inside it.

❺ Plant up the top of the basket with the remaining strawberry and verbena plants and the antirrhinums, placing most of the latter in the centre and towards the back. Fill with compost, pressing it between and around the plants. Add a final thin layer of compost until it is 2.5cm (1in) below the rim and press down firmly. Water thoroughly.

Aftercare

Apply a liquid feed weekly. Pinch out the antirrhinums when they reach 7.5cm (3in) to encourage bushiness. When they begin to go to seed, cut off the flower heads just above the secondary flowering shoots. Pinch off unsightly or overlong strawberry suckers.

Professional know-how ➤ When threading plants through the side of a container, wrap them in a cone of plastic to protect them from damage.

Line the basket with moss, then plastic, and part fill with compost

Wrap delicate plants in plastic to form cones with the leaves at the point

Thread cones through the side of the basket, working from the inside

Fill with compost and plant the strawberries around the top

Add antirrhinums to the centre of the basket for height

Water generously to settle the plant roots into the compost

sugar almond shopper

You will need

12 mixed border primulas
  in pink, such as *Primula
  sieboldii, P. x pubescens
  'Mrs J H Wilson', P.
  whitei 'Sherriff's Variety'*
  (syn. *P. bhutanica*),
  P. edgeworthii, and
  border varieties of
  P. auricula
12 *Campanula
  cochleariifolia*
Slatted shopping basket,
  about 20in (50cm) long,
  9in (23cm) wide and
  deep
Yachting rope or sash
  cord, 3m (10ft)
Sphagnum moss, to line
  the basket
Plastic sheet
Scissors
Broken-up expanded
  polystyrene, such as old
  plant trays
Compost

Season Late spring
When to plant Early to
  mid-spring
Site Light shade

Even the most ordinary baskets around the house can be quickly converted into the hanging variety with the addition of some strong ropes. This pleasing shopper in English willow is a perfect candidate because its strutted construction allows you to thread a few of the plants through the side for a wonderful cascade of flowers that appears to break over the rim of the basket. By crowding plenty of smaller blooms in a large container you can give the impression of a tiny section of flower border, making an English hanging garden in miniature. Here, a combination of border primulas and trumpet-flowered campanulas, in delicate pinks and lilacs, makes a delightful late spring display.

1 Water the plants thoroughly and allow them to drain. Line the basket with sphagnum moss then with plastic sheet. Cut slits in the plastic for side planting. Fill the bottom of the basket with broken-up expanded polystyrene for lighweight drainage and top this with a thick layer of compost.

2 Wrap a primula plant in a square of plastic to form a cone with the flowers in the pointed end. From inside the basket, thread the pointed end through the struts, easing the flowers through carefully. Once they are through, gently pull away the plastic, leaving the plant in position. Repeat with several more primulas.

3 Fill the basket with compost to within about 15cm (6in) of the rim. Plant the remaining primulas and the campanulas in the top of the basket, pressing more

compost in and around the plants until it is 2.5cm (1in) below the rim. Press down firmly then water well.

4 Cut the rope or cord into four sections. Tie one end of each to the basket and the free ends together. Hang from a strong metal hook or bracket.

Aftercare

Both campanulas and primulas like well-drained soil, so do not overwater the basket. Water when the top of the compost is dry to the touch and apply a liquid feed weekly. Deadhead the flowers regularly. When the plants have finished blooming, divide them and plant the resulting plantlets in a frost-free area of the garden or in a greenhouse to protect them from frost. The plants can be used for a new arrangement the following year and spare plants can be used in a border.

hearts & flowers

The flamboyant combination of *Dicentra spectabalis* and crimson tulips is almost a celebration of spring itself, with the delicate heart-shaped flowers of the dicentra symbolizing nature's own season of love. The tulips add colour impact to the whole arrangement, which should give a full month of elegant colour. Choose late-flowering tulips, such as *Tulipa* 'Palestrina', or, for a softer, supremely feminine look, *T*. 'Angélique'. Plant the tulips in pots in late autumn, ready to plant up the basket in mid-spring when they will be in full leaf and about to produce flower buds.

You will need

1 *Dicentra spectabilis*
6 *Tulipa* 'Palestrina' or
 'Angélique', plants
Wire basket, about 25cm
 (10in) diameter
Pot or bucket to stand
 basket on while working
Sphagnum moss, to line
 the basket
Plastic sheet
Compost

Season Late spring to early
 summer
When to plant Late spring
Site Light shade

❶ Water the plants thoroughly and allow them to drain. Stand the basket on a bucket or pot to keep it stable while working. Line it with sphagnum moss, and then with black plastic, slit for drainage and cut to fit. Place a 5cm (2in) deep layer of compost in the basket.

❷ Position the dicentra centrally in the basket then arrange the tulip plants in a circle around it. Take care to plant the tulips at exactly the same depth as they were in the pots before being moved.

❸ Fill the basket with compost, packing it between and around each plant and pressing it it down firmly with your fingers.

❹ Sprinkle the top of the compost with moss. Water the basket well, until water flows from the bottom.

ice cream sundae

A cascade of summer pinks that produce bloom after bloom makes this a highly successful combination for a basket. Other small fuchsias may be substituted for the ones named here.

You will need

8 *Diascia barberae* 'Ruby
  Field'

4 *Verbena* 'Pink Parfait'

1 *Fuchsia* 'Swingtime'

1 *Fuchsia* 'Pink Galore'

1 *Fuchsia* 'Come Dancing'

3 *Fragaria vesca*
  'Semperflorens' (Alpine
  strawberry) plants

2 *Petunia* 'Falcon Red
  Morn'

Hanging basket, about
  35cm (14in) diameter

Pot or bucket to stand
  the basket on while
  working

Coconut fibre liner to fit

Compost

Scissors

Sheet plastic

Season Summer

When to plant Late spring

Site Sunny

❶ Water all the plants thoroughly. Stand the basket on a pot or bucket and fit the coconut liner in place. Cover the bottom 5cm (2in) of the basket with compost.

❷ Make eight planting holes through the liner at equal intervals around the basket and plant the diascias as described on page 22, steps 2 and 3.

❸ Make four more planting holes around the basket and plant the verbenas in the same way.

❹ Add more compost until it is the correct height for the largest of the remaining plants. Test this by resting the pot on the compost in the basket. When the soil in the pot is about 2.5cm (1in) below the rim of the basket, that is the correct amount of compost.

❺ Position the three fuchsias centrally and arrange the strawberry plants and petunias around them. Bed the plants in by adding more compost between and around each plant until it reaches 2.5cm (1in) below the rim of the basket. Press it down firmly.

❻ Water thoroughly, using a fine rose.

Aftercare

Pinch out the fuchsia growing tips initially to encourage bushy growth. During the flowering season apply a liquid feed every four days. Deadhead regularly.

PINK LADIES

Heathers and cyclamen make a pretty autumn planting scheme in an unexpected pink. These plants look charming in an old pink-painted wash tub and will give colour until the end of the season. The heather, *Calluna vulgaris* 'Annemarie', is particularly delightful with its tightly packed, pale pink buds which deepen in colour as they open to a richer carmine rose. It is set off by pink-flowering *Cyclamen hederifolium* (syn. *C. neapolitanum*) whose pretty marbled leaves lighten the scheme. Half fill the container with broken-up expanded polystyrene, then use ericaceous compost for planting.

To lighten the weight of a large hanging container, such as this adapted bucket, and to save on compost, you can fill the bottom part with broken-up expanded polystyrene. Plants are often sold in polystyrene trays which can be saved for this purpose.

raspberry sorbet

Intrinsically beautiful or unusual hanging baskets can add a sculptural quality to an arrangement. When planning a planting in such a container, it is best to choose plants that will not totally envelop it, and so allow it to be part of the overall design for the whole season. This works very well with the deep bucket used here as its form would still show even if the plants tumbled quite a long way over the side by the end of the summer. The froth of deep pink flowers spilling from the top of the bucket makes for an enchanting arrangement with plenty of texture. Despite the unusual combination, all the plants are easy to grow in containers, offering a long-lasting summer display.

❶ Water the plants thoroughly and allow them to drain. Upturn the bucket and drill drainage holes through the bottom, making sure the bucket is stable as you work. Turn the bucket the right way up and cover the holes with crocks or small stones.

❷ Fill the bucket with compost until it is the depth of the knautia's rootball plus 2.5cm (1in) below the rim.

❸ Place the knautia near the edge of the bucket, then add more compost until it is the correct depth for the smaller plants.

❹ Remove the smaller plants from their pots and arrange them in the bucket around the knautia.

❺ Fill in between and around each plant with compost pressing it down firmly. Water thoroughly.

Aftercare
Water the plants regularly to keep the soil moist but not waterlogged. This may be up to twice a day during extremely hot weather. During the flowering season apply liquid feed every four days. Deadhead all the plants regularly.

fresh yellows

The happiest colour of all, yellow is symbolic of sunshine, bringing cheer whatever the weather, and even nature seems to sense when cheer is most needed. Golden daffodils, immortalized by the English poet, William Wordsworth, herald the beginning of spring and brighter days. Later in the year, as the nights begin to draw in, giant sunflowers make the brightest, most flamboyant splashes of yellow, mimicking the sun itself in an effort, it may seem, to prolong the summer. But yellows can be soft, too, and many can be seen in wild flowers, in species such as miniature violas, primroses and cowslips. Team yellow flowers with lush green foliage, or add even more impact by using bright lime-tinged plants.

golden girl

You will need

1 *Alchemilla mollis*
2 *Lysimachia nummularia*
   'Aurea' (creeping Jenny)
3 *Bidens ferulifolia*
6 *Argyranthemum*
   (syn. *Chrysanthemum*
   frutescens) **'Jamaica**
   Primrose'
Garden container, about
   36cm (14in) diameter
Crocks or small stones
Compost

Season Late spring to early
   autumn
When to plant Mid- to late
   spring
Site Partial sun

This joyous froth of lime-green leaves, with its halo of golden flowers, makes an immensely satisfying planting. It provides colour and interest from spring, when the alchemilla comes to life, right through into autumn, when the plants die down with the the arrival of the first frosts. The container will look good as soon as it is planted, but within a few weeks, the lysimachias will have grown into a vibrant cascade of lush lime-coloured leaves and the alchemilla will have thrown up its flirty green flowers.

❶ Water all the plants well in their pots and allow them to drain. Place crocks or stones over the holes in the bottom of the container. Cover these with a 2.5cm (1in) layer of compost.

❷ Gently remove the alchemilla from its pot, tapping it if necessary, and position it in the centre of the container. Fill the container with compost up to a level suitable for the remainder of the plants. This will be the depth of their rootballs plus 2.5cm (1in) below the rim of the container.

❸ Remove the remaining plants from their pots and arrange them around the alchemilla.

❹ Fill in and around all the plants with compost, pressing down firmly, until it is within 2.5cm (1in) of the rim of the container. Water thoroughly taking care not to disturb the plants.

Aftercare
Water well whenever the compost is dry to the touch; in hot weather, this could be twice a day. Feed the plants once a week. Deadhead the bidens and argyranthemums regularly. At the end of the season, cut the plants back hard. If the alchemilla looks as if it has outgrown the pot, take it out and divide it, either in the autumn or in the spring, and replant a smaller section ready for the next season.

Professional know-how ➤ Always leave a 2.5cm (1in) space between the top of the compost and the rim of the container in any planting to allow for correct watering. This space should be filled when watering and the water allowed to drain away through the bottom of the container.

Gather the ingredients together and put some compost in the container

Use a plant in its pot to check that the compost is at the right level

Position the alchemilla centrally and add more compost all round

Surround the first plant with smaller plants, building up the compost to their level

Fill in and around the plants with compost, pressing it down firmly with your fingertips

Water thoroughly but gently, using an upturned fine rose, until water runs from the bottom of the container

lemon syllabub

You will need

2 *Mimulus* Malibu Series,
  yellow
6 *Viola* 'Magnum Cream'
6 *Viola* 'Sorbet Lemon
  Chiffon'
Wire basket, about 46cm
  (18in) long
Sphagnum moss
Plastic sheet
Compost

Season Summer
When to plant Late spring
Site Sunny

Turn a simple, easy-grow combination into something different by using an old functional wire basket, rather than a purpose-made hanging flower basket. To retain the structure of the wire potato picker used here all through the season, the pansies and toning mimulus have been planted into the top of the basket only, and not into the sides. Pansies really are the easiest and most satisfying container plants to grow as they produce flush after flush of blooms all through the season. But they do need attention as they will only do this if you meticulously deadhead, preferably every day.

1 Water all the plants well and allow them to drain. Line the basket with a layer of sphagnum moss, and then with plastic sheet. Make holes in the bottom of the plastic for drainage.

2 Place a layer of compost in the basket and arrange the mimulus plants in the basket. Add compost around and between the mimulus plants until it is the depth of the 'Magnum Cream' viola rootballs plus 2.5cm (1in) below the rim of the basket.

3 Position the 'Magnum Cream' violas around and between the mimulus, distributing them evenly around the basket. Add more compost, if necessary, so it is the

correct depth for planting the small violas.

4 Distribute the 'Sorbet Lemon Chiffon' violas between all the other plants, packing them in well.

5 Fill the basket with compost, pressing it around and between all the plants, then add a final layer to cover, up to 2.5cm (1in) below the rim. Water thoroughly.

Aftercare
Keep the soil moist, watering up to twice a day on hot days. Feed with liquid fertilizer once a week. Check the flowers daily and deadhead whenever necessary to encourage repeat flowering.

a crowd of golden daffodils

Daffodils look best en masse so in small gardens especially, where there is not much space, simply pack them close together in a container. As there is nothing subtle about daffodils, they are best set off by simple containers. This old crate, for instance, offers the perfect solution, giving plenty of planting space. The mixed varieties used lend interesting texture, and by planting in layers you are rewarded with a dense lush effect.

Professional know-how ➤ When planting bulbs in the ground, it is usual practice to place them at twice the depth of the bulbs' height. Use this formula as far as possible when planting bulbs in a container.

You will need

10 *Narcissus* 'Kingscourt', or other large daffodil, bulbs

10 *Narcissus* 'Spellbinder', bulbs

10 *Narcissus* 'Jumblie', bulbs

Wooden box, about 30 x 60cm (12 x 24in)

Coarse gravel

Bulb fibre or compost

Moss

Medium-gauge florist's or garden wire

Season Spring

When to plant Late summer

Site Sheltered

❶ Put a 5cm (2in) layer of gravel in the bottom of the box for drainage. Cover this with a 2.5cm (1in) layer of bulb fibre or compost. Arrange the *Narcissus* 'Kingscourt' bulbs, or other large spring-flowering daffodils if you are using them, in a layer on the bulb fibre. Place the bulbs, with their points upwards, as close together as possible without allowing them to touch each other or the side of the container.

❷ Cover the bulbs with a layer of bulb fibre, leaving their tips just protruding. Place the smaller bulbs (N. 'Spellbinder' and N. 'Jumblie') on the bulb fibre, between the tips of the bulbs below.

❸ Fill the box with bulb fibre to within 2.5cm (1in) of the top then cover this completely with moss. Water thoroughly, then place in a cold greenhouse or a sheltered spot outside, keeping the compost moist until the shoots appear.

❹ As soon as the shoots emerge, fix the moss in place with short lengths of wire bent hairpin style, taking care not to pierce the bulbs below.

Aftercare

When the shoots appear feed every 10 days and move to a sheltered spot outside. When the flowers have died, place in a sunny, secluded spot until the leaves die down. An occasional foliar feed before they die will help strengthen the bulbs.

Cover the bottom of the box with a layer of gravel

Add a layer of bulb fibre and position the large bulbs

Add more bulb fibre, then position the smaller bulbs

Cover with bulb fibre then moss and water thoroughly

hot & sunny

5 *Helianthus annuus*
  'Teddy Bear', or other
  dwarf sunflowers
4 Chilli pepper plants,
  yellow
6 *Viola* 'Sunny Boy', or
  other yellow pansy
Terracotta window box,
  about 40cm (16in) long
Crocks or small stones
Compost

Season Autumn
When to plant Late
  summer
Site Sunny

Let a row of sunflowers bring cheer to your window as the summer draws to a close. At this time of year, dwarf sunflowers, with their extrovert bright yellow flowers, are easy to find on sale in markets and garden centres, providing instant colour for a window box or garden container placed near the door. Hot them up with bright yellow chillis and add charm with golden pansies. The finished container will need plenty of water for these are thirsty plants, but you will be rewarded with colour right through autumn in return for very little effort on your part.

❶ Water all the plants well in their pots and allow them to drain. Place crocks or stones over the drainage holes in the window box and cover the bottom with a 5cm (2in) layer of compost.

❷ Position the sunflowers towards the back of the window box in a row. Add a 2.5cm (1in) layer of compost to the container, packing it firmly around each plant as you go.

❸ Arrange the chilli plants in front of the sunflowers, then add more compost, again pressing it down between the plants.

❹ Finally, arrange the pansies at the front of the window box and fill with compost to 2.5cm (1in) below the rim. Press the compost down firmly all over, between and around each plant to bed them in well. Water thoroughly.

Aftercare
Water the plants regularly and apply a feed once a week. Deadhead the pansies regularly. Once the sunflowers are past their best, cut them right down. The pansies and chillis will have filled out by then and should continue to provide colour for a few weeks more until the first frosts.

First, put in a row of sunflowers

Add a line of chilli plants, and finally the pansies in front of these

PRETTY MAIDS

Marigolds (*Calendula officinalis*) are the most rewarding of summer flowers, producing bloom after bloom for months in return for regular watering and a weekly feed. They are so distinctive yet they need little fussing. So, instead of mixing them with other plants, plant up a trio of coloured pots and set them on a sunny windowsill. Deadhead regularly. Water only when the top of the compost is dry to the touch and do not overwater.

LEMON SOUFFLE below

Crocuses have a delicate charm that belies their resilience. These potted bulbs can survive snow and frost, bringing floral cheer during the coldest days of late winter. As well as *Crocus crysanthus* 'Cream Beauty', the front pot also contains a variegated thyme, which produces evergreen leaves.

TUBULAR BELLES above

Foxgloves (*Digitalis grandiflora*) look enchanting growing in a pot, and this is an excellent solution for small gardens. Combine them with antirrhinums such as *Antirrhinum* 'Liberty White', which will bloom for longer than the foxgloves and so prolong the life of the planting.

DANCING PARTNERS

Capture a springtime miniature woodland scene by combining delicate lime-green ferns with primroses in a wall-mounted pot that can be enjoyed at eye level.

BUCKETS FULL OF SPRING

Hoist spring flowers high up in buckets to bring colour to trees that are still in bud. These lily-flowered tulips (*Tulipa* 'Wespoint') make an elegant combination with primroses.

moody blues

Blue is an extremely versatile colour for planting. Its variations range from the clear tones of forget-me-nots and delphiniums to the deep blues and purples of lavender and pale pinky-lilac shades. When planting in blue, keep to one end of the spectrum, perhaps combining either soft lilacs with purples or the clearer tones of cobalt with royal blue. Alternatively, you could create a tapestry effect by mixing all the blues with abandon. The key to success is to make sure you use at least three shades. Two from opposite ends of the blue spectrum would look like a mismatch, but three or more begin to harmonize, lending depth to the overall scheme.

FLOWER TREE
Display heartsease in a charming new way by planting it up in a cone shape and standing this in an old terracotta pot to create a floral topiary that becomes fuller and fuller as the season progresses.

flower tree

You will need

60 *Viola tricolor*, plant
    plugs
Chickenwire, 1m (1yd) of
    2.5cm (1in) mesh
Wire cutters
Heavy gauge garden wire,
    35cm (14in)
Sphagnum moss
Slow-release fertilizer
    granules
Compost
Terracotta pot, about
    25cm (10in) diameter
    and high
Crocks
Gravel or shingle
Dibber

Season Summer to early
    autumn
When to plant Mid-spring
Site Sun or shade

Display miniature violas in a novel way by creating a potted topiary. The dainty nodding heads of violas take on a whole new look when gathered together in a compost-filled chickenwire cone to make a floral topiary tree. This is not difficult to do, and makes a charming decorative feature for the front window sill or step which will fill out and bloom all summer and right into autumn. Just push one plant plug into each chickenwire 'hole' from the outside, then sit back and wait for a spring flourish. When the violas are past their best, the cone could be emptied out, refilled with fresh compost, and then planted up with grape hyacinth (*Muscari*) bulbs.

❶ Water all the viola plugs thoroughly and allow them to drain. Make a cone from chickenwire. The cone should be about 5cm (2in) taller than the terracotta pot and will need to fit snugly inside it with its base sitting about 5cm (2in) below the rim. Roll the chickenwire roughly into shape, allowing an extra 12cm (5in) at the bottom of the cone. Cut the chickenwire to size, then bend it into shape. Cut the garden wire to fit the pot's internal circumference 5cm (2in) below the rim plus enough at each end to turn back into a hook. Fit this around the outside of the chickenwire cone 12cm (5in) from the bottom and link the hooks to fix the circle. Fix the 'seam' of the cone by bending the cut edges in on each other.

❷ Line the cone with moss. Mix a handful of slow-release fertilizer granules into the compost, then fill the cone with the mixture. Pack the compost tightly into

the cone then bend the extra 12cm (5in) of chickenwire at the bottom inwards to hold the compost in place. Place a crock over the drainage hole of the terracotta pot and fill it with gravel or shingle to within 5cm (2in) of the rim, then fit the cone in the pot.

❸ Carefully make an angled planting hole through the moss and into the compost, using a dibber or your finger, and gently push in a viola plug. Plant violas all round the cone in this way to cover it evenly. At this stage, some moss will still be visible between the plants, but within two to three weeks the plants should have grown sufficiently to completely cover all the moss. Water thoroughly.

Aftercare
Water the cone frequently, using a watering can fitted with a fine rose. Deadhead the flowers regularly.

Fashion chickenwire into a cone shape

Line with moss and fill with compost. Bend over the chickenwire to close

Place the cone upright in the pot

Plant up the cone through the mesh

Professional know-how
➤ This cone should be planted up with plant plugs when their rootballs are still small enough to fit through the chickenwire mesh.

purple haze

You will need

1 ornamental kale
2 *Ajuga reptans*
    'Multicolor'
3 *Aster novi-belgii*
    'Professor Anton
    Kippenberg'
Galvanized metal hanging
    tray, about 30cm (12in)
    diameter and at least
    7.5cm (3in) deep
Electric drill with metal
    bit
Crocks or small stones
Shingle
Compost

Season Autumn
When to plant Very late
    summer or early autumn
Site Full sun

The mists of autumn bring with them all the soft amethyst shades that happily mix together while combining well with seasonal blue-green foliage. The shapes tend to be more structured and slow-growing than those of summer annuals, gently filling out as the season progresses. As summer fades, ornamental cabbages come into season, offering a choice of wonderful sculptural shapes in bluey greens and whites, often shot with purple. Just one of these makes a fabulous focal point for an autumn hanging basket. Here, such a cabbage has been teamed with the wonderful crimson foliage tones of a bugle (*Ajuga reptans* 'Multicolor') and the soft lavender shades of *Aster novi-belgii* 'Professor Anton Kippenberg', which has a compact bushy habit suitable for containers. The soft grey of galvanized metal perfectly complements these pretty purple shades.

❶ Water all the plants thoroughly and allow them to drain. Using an electric drill, make drainage holes through the bottom of the tray. Make sure the tray is steady before you start to drill.

❷ Cover the drainage holes with crocks or small stones then line the bottom of the tray with shingle to aid drainage. Top this with a 5cm (2in) layer of compost.

❸ Position the kale towards one side of the tray. Place one bugle plant to the front and the other to the back. Arrange the asters between these.

❹ Press compost around each plant, packing it firmly. Fill the tray with compost to within 2.5cm (1in) of the rim and press it down firmly. Water the container thoroughly.

Aftercare
Water whenever the top of the soil feels dry to the touch. Apply liquid feed once a week. Deadhead the asters regularly and when they finish flowering completely, cut down the stems to the level of the compost. In spring, divide both the aster and bugle plants, replanting some plantlets in the tray and distributing the remainder around the garden.

Place a layer of shingle, then a layer of compost in the tray

Plant the kale, bugles and asters

blue lagoon

You will need

10 *Petunia* Surfinia® 'Blue
   Vein'
6 *Convolvulus sabatius*
   (syn. *C. mauritanicus*)
10 *Lobelia* 'Kathleen
   Mallard'
1 *Verbena* 'Aphrodite'
3 *Nemesia caerulea* 'Joan
   Wilder'
2 *Viola tricolor*
   (heartsease, wild pansy)
1 *Viola* 'Universal Series',
   ivory/rose blotch
**Window box, about 60cm
   (2ft) long**
Electric drill (optional)
Crocks or small stones
Compost

Season Summer to early
   autumn
When to plant Mid-spring
Site Sun or shade

Mix blues in many fast-growing varieties for a richly textured
effect to create a lush window box that becomes fuller as the year
progresses. This container has been filled with fast-growing
plants that produce an abundance of blooms which, as long they
are deadheaded regularly, will keep on coming right through the
summer. Using two window boxes – one on a window sill and
the other on brackets just in front of it – can double the effect.

❶ Water the plants thoroughly and allow them to drain. Drill drainage holes
through the bottom of the container, if necessary. Cover the drainage holes with
crocks or small stones then put a 15cm (6in) layer of compost in the container.

❷ Remove the plants from their pots. Position the petunias at the back of the
window box and the convolvulus and lobelias at the front, then arrange the
remaining plants between them.

❸ Fill the window box with compost packing it around and between the plants and
pressing it down firmly.

❹ Water the container thoroughly.

Aftercare
Water the window box whenever the surface of the compost feels dry to the touch;
this could be up to twice a day in the heat of the summer. Apply a liquid feed once a
week. Deadhead the petunias and violas daily to promote flowering. Prune the
verbena and lobelia heads when the flowers have died.

BRIGHT AND BEAUTIFUL

Sometimes, single plantings can be spectacular. This exotic *Scaevola aemula*, native to Australia, is new to our shores. It produces glorious blue trailing flowers all through the summer and, if overwintered in a greenhouse or conservatory, it will provide the same show all over again next summer. Scaevolas grown in containers need a lot of watering.

GROWING UP

Standard hydrangeas make glorious summer focal points. This example of *Hydrangea macrophylla* has been underplanted with toning *Nepeta racemosa*, which lends a gentle frothy appearance to the top of the pot. Hydrangeas are one of the many plants that can now be bought already shaped into standards. Preserve its outline by pruning out any low horizontal shoots that develop. Water thoroughly. If the hydrangea has a tendency to revert to pink, apply hydrangea colourant, once a week when watering. In winter, cut back the dead flowers and leaves, retaining the general shape of the plant.

chicken-in-a-bucket

At times of celebration it is fun to make witty arrangements with plants. This whimsical mossy chicken, sitting on a nest in an enamel bucket planted with muscari, makes an amusing planted container. Place it on a doorstep or window sill as an Easter welcome, or make it up to give as an imaginative gift. If you are unable to obtain a three-dimensional topiary chicken shape, you could use a hen-shaped egg carrier instead. However, there is now a wide variety of three-dimensional wirework topiary shapes available so you could easily adapt this planting idea to suit different animals. A rabbit would make a delightful Easter alternative, for instance, while for the summer you may prefer a peacock.

❶ Water the plants thoroughly and allow them to drain. Drill drainage holes through the bottom of the bucket and cover these with crocks. Fill the bucket with bulb fibre or compost to within 15cm (6in) of the rim.

❷ Line the chicken with moss, then fill its body with bulb fibre or compost to within 10cm (4in) of the top.

❸ Remove the muscari plants from their pots and place them around the inside of the perimeter of the bucket. Press bulb fibre in between and around each plant. Fill the centre of the bucket with bulb fibre to about 2.5cm (1in) below the rim. Place the chicken wire shape in the middle of the bucket.

❹ Plant the iris plants in the chicken's back. Fill any spaces between the plants with compost and press it down firmly. Finish by placing a little carpet moss around the iris stems to cover the bulb fibre or compost. Water thoroughly.

Aftercare
Place the container in a sunny spot. Water whenever the soil feels dry to the touch, but avoid overwatering. When the blooms have died, leave the arrangement in a sunny place until the foliage dies down. Then divide the iris bulbs and store them dry until autumn, when they can be planted out for flowering next spring. The muscari will need dividing every three years.

As a variation on this
Easter theme, sit the
chicken on a nest of forget-
me-nots and plant up with
anemones.

pretty and witty

You will need

2 *Salvia officinalis*
   'Purpurascens'
   (purple sage)
12 *Viola tricolor*
   (heartsease, wild pansy)
5 small terracotta pots
Crocks
Compost
Metal bottle carrier or
   small hanging basket
Garden wire

Season Summer
When to plant Mid-spring
Position Sun or shade

Create a whimsical hanging basket by planting purple sage in a metal bottle carrier, then wiring four terracotta pots planted with tiny violas to it. This delightful idea could be easily extended to any small hanging basket. The combination of flowers used here works particularly well as the sage's young purple tips complement the purple in the violas perfectly.

❶ Water the plants thoroughly and allow them to drain.

❷ Put a crock in the bottom of one of the terracotta pots and add a layer of compost. Remove the sage from its pot and place it in the terracotta pot. Add more compost around the plant and firm it in. Fit the pot into the bottle carrier or basket.

❸ Pass a long piece of garden wire through the hole in the bottom of one of the remaining terracotta pots. Twist the shorter end and the longer section together at the top of the pot then use the longer section to attach the pot to the outside of the bottle holder or basket.

❹ Put a crock then a layer of compost into the pot. Remove three viola plants from their strips or pots and place them in the prepared terracotta pot. Fill in with compost around and between each plant, pressing down firmly.

❺ Wire and plant up the remaining three terracotta pots in the same way. Water all the plants thoroughly.

Aftercare
Deadhead the violas very regularly to promote repeat flowering.

hot & spicy

**COMING UP
ROSES**
This decorative
Victorian-style
basket makes an
elegant container for
delightful single-
bloomed trailing
roses which have
been complemented
by saxifrage to give
spring colour and
red lobelia for
summer fullness.

For impact, nothing can beat the brights. Choose relentless red, piling scarlet on claret, magenta on cherry; or combine the oranges for a fabulous flamed effect. Another way to go hot and spicy is to take inspiration from sun-baked climates, such as Mexico and India, where they clash the colours, using fuchsia pinks with vibrant oranges to great effect. Reds and oranges span all the seasons, showing up on berries in autumn and winter, and flowers in summer and spring. Gathering these vibrant tones together in containers and hanging baskets intensifies their impact, offering focal hot spots around the garden, and even adding instant colour between seasons when it may be lacking.

coming up roses

You will need

2 *Rosa* Suffolk

4 *Saxifraga* x *arendsii*

6 *Lobelia erinus* 'Cascade Red'

Hanging basket, about 35cm (14in) diameter

Pot or bucket to stand basket on while working

Carpet moss

Plastic sheet

Scissors

Compost

Season Spring to summer

When to plant Early spring

Site Sunny

A hanging rose garden makes a delightful summer sight, so enjoy the roses at eye level by planting them in an elegant Victorian-style hanging basket. Choose a deep basket and plant it with a low spreading rose. This one comes from the County Series of repeat-flowering roses with a trailing habit and charming single blooms that are ideal for hanging baskets. Each 'county' is a different colour, to suit endless plantings. When this basket was made up in spring, the rose was underplanted with a mossy saxifrage to provide early interest and with trailing lobelia for additional colour in summer. Given plenty of water and regular applications of liquid feed, the roses in this basket will become fuller and produce longer trails next year.

1 Water all the plants thoroughly and allow them to drain. Stand the basket on a large pot or bucket for support while you work and line it with carpet moss. Line the moss with plastic sheet and pierce this to make drainage holes.

2 Place a 5cm (2in) layer of compost in the basket. If you wish to plant through the sides, make holes and thread some lobelia plants through the sides, following the instructions on page 22, steps 2 and 3.

3 Remove the roses from their pots and position them at opposite sides of the basket, spreading the roots out so they will not be restricted. Add another layer of compost, 5cm (2in) deep, working it around the plants and pressing it down firmly.

4 Place the saxifrage plants between the roses, then position the remaining lobelias between the saxifrages and the rose plants. Fill in with compost, pressing it down firmly. Water thoroughly.

Aftercare

Apply a liquid feed every three weeks. As the roses grow, train their stems down over the sides of the basket, tying them to the rim with soft garden string. Also, train the lobelia through the sides of the basket if, like the one shown here, it has high sides. Deadhead the roses regularly to encourage repeat flowering throughout the season. When the roses stop flowering, trim them to tidy them. Early next spring, untie the rose stems and prune them back by two thirds of the previous summer's growth to promote bushiness.

Place the basket on a pot to steady it, then line the basket with moss

Line the moss with plastic, and put a layer of compost in the bottom

Position the roses on opposite sides

Position the saxifrages and infill with lobelias

Professional know-how
➤ Deadhead roses by cutting back to about 1.5cm (½in) above the second leaf joint below the flower. Slant the cut down and away from the leaf bud. Use sharp secateurs.

in the red

6 small *Hedera helix*, such as 'Babyface, 'Pixie' or 'Mapleleaf'
6 *Verbena* 'Nero'
6 *Petunia* Picotee Series, red
6 *Viola x wittrockiana* 'Floral Dance', or other red pansies
3 *Pelargonium* 'Elsi', or other pink ivy-leaved pelargoniums
Wooden garden trug or basket, about 40cm (16in) long
Plastic sheet
Compost
Sash cord

Season Summer
When to plant Early summer
Site Sun or light shade

At the height of summer clash the colours for a hot, vibrant look that is reminiscent of warm, sunny climes. This wooden garden trug, crammed to overflowing with assorted reds, is spiced up with pinpricks of hot pink pelargoniums for a charming rural look. The variety of crimson blooms in assorted shapes and sizes lends a rich textural quality that will develop a looser, wilder appearance as the season progresses and the plants grow taller and leggier. As summer draws to a close, autumn offers a palette of softer reds which can be used as replacements for the by then unruly petunias. Search out rich crimson asters such as *Aster novi-belgii* 'Royal Ruby'. If they are looking a little tired, the pansies, too, can be replaced with a later-flowering red or white variety. The slatted construction of the trug offers excellent drainage, making it a perfect choice for a hanging basket.

❶ Line the basket with plastic sheet cut to size, and cut two or three slits for drainage.

❷ Place a 5cm (2in) layer of compost in the bottom of the basket. Remove the *Hedera helix* plants from their pots and place them around the edge of the basket.

❸ Add a second layer of compost then arrange the remainder of the plants in the basket to make a pleasing display.

❹ Press additional compost between and around each plant, making sure each one is bedded in well. Add a final topping of compost to reach 2.5cm (1in) below the rim of the basket, and press that down firmly.

❺ Loop a length of sash cord through the basket's handle and tie a knot. Use the cord to hang the basket from a bracket, wedging the side of the basket against the wall for support, if necessary. Water thoroughly.

Aftercare
Water regularly, especially during hot weather. Deadhead all the blooms regularly. If the petunias become too leggy, cut back their stems to about 10cm (4in) above the compost to encourage bushier growth. Cut one plant down at a time, about one week apart, to avoid denuding the basket.

Crammed together in a garden trug, a fiery assortment of red blooms provides a splash of welcome colour against an otherwise somewhat sombre background. Bright flowers are perfect for creating focal hot spots around a garden or for enlivening dull areas such as concrete or paving.

flaming whoppers

You will need

20 *Tulipa* 'Apricot
   Beauty', bulbs
12 *Erysimum*
   (syn. *Cheiranthus*)
   'Orange Bedder'
Large florist's bucket
Drill and metal bit
Broken up expanded
   polystyrene, or other
   filler
Compost

Season Late spring
When to plant Autumn
Position Sun or shade

Orange wallflowers and flame-coloured tulips make a fabulous extrovert combination to herald brighter spring days, with the subtle, sweetly perfumed wallflowers perfectly complementing the flamboyance of the tulips. Enjoy the glorious aroma of the wallflowers daily and at close quarters by placing the container near the front door. Plant the tulip bulbs in autumn. As soon as young wallflowers become available carefully bed them in above the tulip bulbs, then wait for a fiery show!

❶ Drill holes in the bottom of the florist's bucket and half fill it with broken up expanded polystyrene, to about 25cm (10in) below the rim. This will be lighter and cheaper than completely filling the bucket with compost.

❷ Add about 7.5cm (3in) of compost. Place a layer of tulip bulbs on top of this, close together but without touching and with their pointed tips facing upwards.

❸ Cover the bulbs with compost, leaving the tips uncovered. Add another layer of bulbs, placing them between the tips of the bulbs below.

❹ Add a 2.5cm (1in) layer of compost. Arrange the wallflowers on top, taking care not to disturb the bulbs beneath then add more compost, working it around each plant. Press it down well then water the container thoroughly.

Aftercare
Pinch out the wallflower tops to encourage bushiness. When all the flowers have died, place the container in a sunny spot so the bulbs can ripen, and water occasionally. Wallflowers may flower next year, so transfer to another pot to rest.

HOT CHILLI

Capsicums of all kinds provide fun autumn colour, and look best potted up in traditional terracotta shades. Here, a pair of pots planted with different varieties of red hot chilli peppers and supplemented with rust- and orange-coloured pansies are suspended on garden wire to make an unusual hanging display.

BRONZE BEAUTY
Normally prized for
their showy blooms,
pelargoniums can also
show fine foliage.
Pelargonium 'Vancouver
Centennial', for
example, with its
exquisite bronze foliage
makes for a stunning
display when set off by
a delicate copper
hanging basket.

BERRY BRIGHT above
Berries are brilliant winter brighteners.
If your garden is devoid of them, or the
show is less than flamboyant, capture
the look by planting up some in a
hanging basket. This one, hung in a
heavily fruited crab apple (*Malus*) tree,
is planted with *Gaultheria procumbens*.

warm cinnamon

You will need

1 *Rosa* Abraham Darby
1 *Rosa* Sussex
2 *Dahlia coccinea* 'Moonfire'
2 *Lilium* 'Orange Pixie'
1 *Mimulus aurantiacus*
2 *Hemerocallis* 'Toyland'
Galvanized bath or garden container, about 50cm (20in) long
Drill and metal bit
Crocks or small stones
Compost
Hand trowel, optional

Season Summer to early autumn
When to plant Mid-spring
Site Sunny

Mix the cinnamon shades using long-lasting perennials for a container of colour that will last right through summer into autumn. Roses and lilies, dahlias and mimulus, all take turns to provide a bright splash and as blooms on one plant fade, another plant's flowers provide colour while the first starts to produce its next flush of blooms. In this way, the planting seems to make a floral dance all through the season. This arrangement has been planted in an old galvanized bath, its soft grey tones perfectly complementing the loud colours of the blooms, which range from the softest apricot to vibrant orange. Use it as a feature for providing colour impact in the garden throughout the season.

1 Water the plants thoroughly and allow them to drain. Drill drainage holes through the bottom of the galvanized bath.

2 Place a 5cm (2in) layer of crocks or small stones in the bottom of the bath for drainage. Top this with a 15cm (6in) layer of compost.

3 Remove the plants from their pots and arrange them in the container. The tops of the rootballs should be approximately 2.5cm (1in) below the rim of the bath or container. If the rootballs of the larger plants stand proud of the rim, dig holes in the compost deep enough to accommodate each plant. Use a hand trowel to dig the holes.

4 Fill in and around each plant with compost, pressing it down firmly. Water thoroughly.

Aftercare
Water whenever the compost becomes dry to the touch, but do not allow the container to become waterlogged. Apply a liquid feed once a week. Deadhead the roses regularly. In late autumn prune any very long rose stems to avoid their becoming damaged. Cut back all the other plants. Place the container in a frost-free spot, preferably under glass. Continue to water sparingly. In early spring, prune the roses, removing two thirds of the previous year's growth to promote bushiness. Place the container outside once the danger of frost is over.

purest white & green

The gentle elegance of white is a winning solution throughout the year. It complements the snow in winter and lightens summer plantings. White is always tasteful, white never jars. At its best, white is used white on white, with varied textures for an ethereal quality. Set against any colour, green provides the structure of the garden, and usually plays a supporting role. But green can become a feature: in the form of a distinctive tree used as a focal point; as a piece of topiary; or simply as interesting foliage presented in a container to give it impact, then positioned in a prominent part of the garden.

CASCADE
Tumbling down from a large strawberry pot, Surfinia® petunias and trailing pelargoniums put on a pretty summer show. Choose scented pelargoniums, and place the arrangement on a window sill to allow the aroma of summer to waft indoors.

cascade

You will need

3 *Petunia* **Surfinia**®
  '**Kesupite**'

3 *Pelargonium*, **white-**
  flowered varieties as
  available

Strawberry pot, 25cm
  (10in) high

Garden hose, short
  length

Secateurs

Cork, to fit the hose

Crocks or small stones

Compost

Season Summer

When to plant Very late
  spring

Site Sunny

Drifts of petunias, teamed with dainty, delicately scented pelargoniums, tumble from a strawberry pot in a summer-long perfumed profusion of white. Used here is one of the Surfinia® petunias, a most obliging type which produces long elegant trails of blooms as the season progresses in return for very little effort. This type does not become leggy, as do other petunias and so does not need pruning, but it still requires regular deadheading to encourage flowering. Plant some in the sides of the strawberry pot to create a full-skirted effect for an unusual display on a patio or window sill.

1 Water the plants thoroughly and allow them to drain. Cut the hose 5cm (2in) shorter than the pot height and make holes down its length, using secateurs. Stop up one end with a cork.

2 Cover the drainages holes in the strawberry pot with crocks or stones. Place the hose upright in the centre of the pot and fill the pot with compost to 2.5cm (1in) below the lowest planting hole.

3 Remove one of the pelargoniums from its pot and gently push the rootball through one of the lowest planting holes. Repeat on the other side of the pot with another pelargonium.

4 Place a petunia through each of two upper planting holes. Fill the pot three-quarters full with compost.

5 Position the remaining pelargonium and petunia plants in the top of the pot.

6 Fill the pot with compost to within 5cm (2in) of the rim, packing it around the plants and pressing it down firmly. Water thoroughly.

Aftercare
Water when the compost is dry to the touch but do not overwater. Pinch out the growing tips of the pelargoniums to encourage bushiness. Deadhead the pelargoniums and the petunias regularly.

Professional know-how ➤ Strawberry pots can be used to create stunning displays using any trailing plants. Inserting a perforated hose works as a simple irrigation system, by providing a small reservoir of water to keep the compost moist for a longer period of time between waterings.

Cut the hose to length, and make holes along it at intervals

With the hose placed centrally, fill the pot with compost to the lowest hole

Start planting through the side holes, taking care not to damage the roots

Add more compost to fill the pot three-quarters full

Plant the top of the pot with the remaining plants

WHITE MISCHIEF

Create an outdoor 'flower arrangement' for the spring by planting pure white daffodils in a decorative wire basket then, when they have grown high, supporting them with peasticks and tying them around with raffia. These white *Narcissus* 'Thalia', with their delicate trumpets and elegant reflexed petals, make a delightful show. Plant the bulbs in late autumn, first lining the basket with white reindeer moss and sheet plastic. Add more moss to cover the top once the planting is complete. As the bulbs grow, they will lift the moss, so pin it down using short lengths of florist's or garden wire bent hairpin style. By early spring, the leaves should be at least 10cm (4in) high.

WHITE LADIES

Snowdrops, such as these *Galanthus elwesii* 'Flore Pleno' and *G. nivalis* planted into a pair of tiny reindeer moss-lined hanging baskets, make an enchanting feature in the dead of winter. Old chandelier drops lend a New Year celebratory feel while reflecting the soft winter light. The chandelier drops are fixed to the hanging basket using medium-gauge florist's or garden wire.

WHITE MAGIC

Set in moss in a simple wire basket, the only companions these white violets (*Viola sororia* 'Albiflora') need are some scallop shells to dress the top of the compost.

chartreuse

You will need

1 *Helleborus niger*
  (Christmas rose)
6-12 *Anemone blanda*
  'White Splendour'
Small wire window box,
  at least 15cm (6in) deep
  and 20cm (8in) long
Sphagnum moss, to line
  the basket
Plastic sheet
Scissors
Compost
Medium gauge florists'
  wire or garden wire

Season Early winter
When to plant Late spring
Site Light shade

Deliciously light and airy, the combination of a hellebore and *Anemone blanda* makes for the most original and delightful winter window box. Ideally, this arrangement should be planted in the autumn when the anemone tubers are dormant, as these plants object to root disturbance. Since the tubers sometimes do not shoot, it may be wise to plant up double the numbers required.

1 Water the hellebore thoroughly and allow it to drain. Line the wire window box with sphagnum moss, and then with plastic sheet cut to fit. Place a 5cm (2in) layer of compost in the bottom.

2 Place the hellebore to one side of the box and position the anemone tubers around it.

3 Fill the window box with compost to within 2.5cm (1in) of the rim, packing it around the plants and pressing it down firmly.

4 Cover the compost with a layer of sphagnum moss and fix it in place with short pieces of wire bent hairpin style. Water thoroughly.

Aftercare
Keep the compost moist. When the hellebore has finished flowering, cut the flower stems off and plant it out in the garden as it will grow too big for this small window box. The anemones can either be left in the box or planted out.

herb basket

A basket of herbs is both pretty and functional, as the summer growth will provide plenty for the cooking pot while making a decorative and aromatic feature outside the kitchen window. This one, packed mainly with culinary herbs, has been given a decorative touch with the addition of lavender, violas and a 'necktie' of raffia. Herbs are obliging plants, many being easy to look after as they require little in the way of water or fertilizer. However, within the confines of a container, they will need a little extra attention. The mint, for example, may need regular trimming to prevent it taking over.

❶ Water all the plants thoroughly and allow them to drain. Line the bottom 7.5cm (3in) of the basket with moss and then put a disc of plastic at the bottom of the basket to act as a water reservoir.

❷ Place a 7.5cm (3in) layer of compost in the basket.

❸ Put six of the upright-growing plants, such as the purple sage (*Salvia officinalis*), lavender, hyssop, rosemary and strawberry, to one side then divide the remaining herbs into two mixed groups ready to plant them through the sides of the basket in two layers.

❹ Plant the group which includes the pineapple mint first. Lay the mint diagonally on a square of plastic sheet and roll it up to form a cone with the point at the foliage end. Make a small hole through the moss with your finger. Thread the plastic cone through this hole from inside the basket, pointed end first, until the rootball is firmly against the moss. Gently pull away the plastic leaving the plant in position. Plant the remaining plants of the group at intervals around the basket.

❺ Line the basket with another 7.5cm (3in) high band of moss and fill this with more compost. Plant the second group of plants through the side of the basket as described in step 4.

❻ Fill the basket with compost to within 7.5cm (3in) of the rim. Position the reserved plants in the top of the basket and fill in with compost, packing it around the plants and pressing down firmly. Water thoroughly.

Aftercare
Water when the top of the compost feels dry. Pinch off the tips of the herbs regularly to prevent them flowering and cut off leaves as you need them. Cut down the lavender flower stems at the end of autumn.

FIG CONFECTION

It is surprising what you can grow in a pot. This fig, *Ficus carica* 'Brown Turkey', in an oriental ceramic pot, perfectly complements its larger-growing cousin which is planted in the ground. Potting plants from warmer climates is a good idea where there is a threat of winter frost as the whole ensemble can be brought into the conservatory or greenhouse for over-wintering.

GREEN AND PLEASANT below

Decorative cabbages (kale), ranging from these full traditional shapes to wispy, frondy varieties, all in glorious shades of green and purple, make fabulous autumn and winter plantings. They are full and verdant, and keep their shape right through the season. The colours work brilliantly with galvanized metal garden containers – all you need to add is a little vibrant carpet moss.

UNDERNEATH THE GOOSEBERRY BUSH

This enchanting almost fairy-tale tree is actually an ordinary garden gooseberry that has been trained into a standard. It was grafted onto a 1m (39in) stem by the nursery, and then kept trimmed in shape. Underplanting with heathers keeps the interest going at pot level. These can be treated as bedding plants, and changed with the season to keep the arrangement looking fresh.

round and round the garden

In winter, most gardens would benefit from some evergreen structure. The quickest and simplest way to do this is to introduce a clipped evergreen. This spiral × *Cupressocyparis leylandii* 'Castlewellan' makes a striking feature when underplanted with white heather, cheering a cold winter garden. Finding a large enough container can prove a problem, which is solved inexpensively here by using a half barrel, brightened up with a simple stencilled design inspired by Elizabethan strapwork.

Professional know-how ➤ Stencils in various designs can be bought from a craft shop, but you can make up your own designs. Draw the outline onto waxed stencil paper, lay it on a cutting mat, then cut it out using a scalpel.

You will need

Large half barrel

Gloss paint: turquoise, royal blue, yellow ochre, burnt orange and dark blue

Small decorator's paintbrushes

Masking tape

Waxed stencils: fleur-de-lis and strapwork

Saucer

Stencil brush

Newspaper

1 Coat the whole of the outside, the top edge and the top 10cm (4in) of the inside of the barrel with turquoise paint. Allow to dry completely.

2 Mark off vertical stripes for the strapwork with masking tape so that they narrow from top to bottom. These stripes are 9.5cm (3¾in) wide to accommodate the strapwork and 15cm (6in) apart at the bottom to allow the fleur-de-lis motif to fit between them.

3 Paint the barrel up to and including the top metal band in royal blue, leaving the sections between the masking tape turquoise. Allow to dry.

4 Peel off the masking tape. Fix the fleur-de-lis stencil in position over a royal blue stripe with tape. Pour a little yellow paint into a saucer, dip the end of the stencil brush in it and work off the excess paint on a piece of newspaper. Dab the paint onto the stencil area. Unless you are very skilled, it is best to allow the paint to dry before removing the stencil and stencilling another motif. Alternatively buy or cut more stencils and leave them in place so that you can paint all the motifs at the same time.

5 Stencil the strapwork motif on the turquoise panels as described in step 4.

6 Paint the wood above the top metal band in alternate stripes of yellow and burnt orange. Highlight the strapwork with dark blue 'shadow'.

Paint the barrel all over in turquoise

Paint on royal blue stripes, using masking tape guides

Stencil the fleur-de-lis motif onto the royal blue stripes

Stencil the strapwork motif onto the turquoise panels

sepia shades

There is something very beautiful about the soft shades that are reminiscent of early sepia photographs. Spanning muted old golds to pinky apricots, these have become favourite colours in recent years. Plants such as some houseleeks (*Sempervivum*) and hellebores have always come in these delicate tones. Now, roses, foxgloves and verbascum are joining the ranks, offering a varied palette in many varieties that can be put together with bronze foliage to create very unusual containers. These gentle shades are available right through the year, with hellebores spanning winter and early spring.

ROCKY ROAD
Gathered together in large terracotta pans, alpines in pots within pots make an exquisite display, their soft sepia tones perfectly complementing their containers. Not difficult to look after, this arrangement will give pleasure year after year.

rocky road

You will need

1 *Sempervivum* 'Polaris'
1 *Sempervivum* 'Tenbury'
1 *Sempervivum* 'Blue Boy'
1 *Sempervivum* 'Jungle
  Fires'
1 *Delphinium nudicaule*
1 *Mimulus* 'Highland
  Orange'
1 *Lewisia cotyledon*
1 *Ajuga reptans* 'Rainbow'
5 terracotta pots,
  5cm (2in) diameter
7 terracotta pots,
  10cm (4in) diameter
1 terracotta pot,
  10cm (4in) diameter
  with a chipped edge
Large alpine pan, about
  50cm (20in) diameter
Alpine compost
Fine gravel
Pebbles
Crocks
Compost

Season All year
When to plant Early
  autumn or spring
Site Sunny

Dainty alpines deserve a close look, so they are best displayed almost as an exhibit. Here, eight different types are shown off, each in its own pot set within a large shallow alpine pan to provide a wonderful combination of sepia shades. Once planted, they will thrive and grow with little attention apart from occasional watering. One advantage of planting pots within pots in mixed plantings is that if, for any reason, one plant does not survive, the pot can be easily exchanged for another, newly planted one. The evergreen houseleeks (*Sempervivum*) will spread over the years forming a mat, and in the summer throw up exquisite pink star-like flowers on elongated stalks. If the houseleek varieties specified here are unavailable, any others can be substituted.

❶ Water all the plants and allow them to drain. Pot up each sempervivum in a 5cm (2in) pot, using specially formulated alpine compost. Place some gravel in one of 10cm (4in) pots, then stand the small planted-up pot in it and fill between the pots with gravel. Fill three more 10cm (4in) pots in the same way.

❷ Cover the bottom of the alpine pan with a layer of pebbles and arrange the potted-up sempervivum plants all around the edge.

❸ Pot up all the remaining plants, except the lewisia, in individual 10cm (4in) pots. Do this by putting a crock over the drainage hole then adding a layer of compost. Place a plant in the pot and fill around it with more compost, pressing down firmly. Cover the compost with a layer of gravel.

❹ Plant the lewisia in the broken pot so that it will grow upwards when the pot is laid on its side. Cover the compost with a layer of gravel.

❺ Arrange the 10cm (4in) pots in the middle of the pan, wedging the broken pot securely at an angle.

❻ Fill the pan and around all the pots with gravel and add some pebbles for decoration.

Aftercare
Most alpines dislike being too wet, so take care not to overwater them. The lewisia does not tolerate water in its crown, so always water it beneath its leaves. Deadhead the mimulus and delphinium as necessary.

Professional know-how ➤ These mountain-dwelling plants like well-drained compost and if they get too wet, cushion alpines especially, will rot at the neck. A top dressing of grit stops the compost from splashing up during wet weather.

Collect together all the different ingredients for the planting

Pot up the sempervivums in tiny pots. Place each in a larger pot with gravel

Stand the sempervivums around the edge of the pan on a layer of pebbles

Pot up the remaining plants in individual pots, putting the lewisia in the broken one

Arrange the plants in the middle of the pan, wedging the lewisia to one side

Add gravel in and around all the pots, then add flat pebbles for decoration

black & tan

You will need

- 1 *Euphorbia griffithii* **'Fireglow'**
- 1 *Trifolium repens* **'Purpurascens'**
- 3 *Viola* **'Molly Sanderson'**
- 1 *Coleus blumei* (syn. *Solenostemon*) **(flame nettle)**
- 3 *Viola* **Universal Series, orange blotch**
- 2 *Papaver rhoeas*, **orange (field poppies)**
- **Terracotta pot, about 35cm (14in) diameter**
- **Crocks or small stones**
- **Compost**

Season Spring to summer
When to plant Mid-spring
Site Sunny

Here is a stunning container that is a far cry from convention, yet very easy to grow and keep. Anyone who wants something different will be more than delighted by this imaginative yet highly successful grouping which has been planned to give a sophisticated black and tan combination from spring right through the summer. The euphorbia provides a brilliant orange show in spring, and as its colour dies, similarly shaded poppies come into bloom. Rust-coloured coleus and pansies also keep up the tan momentum. All the while, black clover and tiny black violas provide a fascinating rich background. Growing fuller as the season progress, this planting nevertheless retains a smart overall look. Keep it that way by meticulously deadheading the pansies every day.

❶ Water all the plants thoroughly and allow them to drain. Place a crock or small stones over the drainage hole of the terracotta pot and fill with compost until it is the correct height for planting the euphorbia. Rest the euphorbia pot on the compost. When the top of the pot's compost is 2.5cm (1in) below the rim of the container, the compost is the correct depth.

❷ Remove the euphorbia from its pot and place it towards the back of the container. Fill in with compost until it is the correct height for the remaining plants.

❸ Remove the remaining plants from their pots and arrange them in the container.

❹ Fill in between and around all the plants with compost, pressing it down firmly as you go. Water thoroughly.

Aftercare

Water thoroughly whenever the top of the compost becomes dry to the touch. Apply a liquid feed weekly. Deadhead the pansies when necessary, which could be every day at times; this will promote repeat flowering. Cut the poppies down to compost level when they have finished blooming.

copper & candy

You will need

1 *Heuchera micrantha
    diversifolia* 'Palace
    Purple'
1 Ornamental purple
    kale, feathery leaves
2 *Bellis perennis* (daisy)
1 *Ajuga reptans* 'Burgundy
    Glow'
**Painted galvanized
    container**
Drill and metal bit
Crocks or small stones
Compost

Season Autumn to winter
When to plant Very late
    summer or early autumn
Site Partial sun

Teaming copper and bronze with the brightest of pinks achieves the look of an old, hand-tinted sepia photograph. Although these are the normal shades of these plants, they make a glorious combination for autumn, when all around greens are turning to bronze. At the same time, the daisies offer a surprisingly summery feel, and the whole container can be placed in a spot where summer bedding is past its best and the borders need reviving. The rest of the plants are evergreen, creating an interesting combination that will go on right through winter, the feathery leaves of the kale offering a delicate contrast to the glossy-leaved heuchera and ajuga.

1 Water all the plants thoroughly and allow them to drain. Drill drainage holes through the bottom of the container, if necessary, using an electric drill fitted with a metal bit.

2 Place crocks or small stones over the holes then cover the bottom of the container with a layer of compost at least 5cm (2in) deep.

3 Place the heuchera and the kale in their pots on the compost. Adjust the level of compost so that the top of the compost in the pots is 2.5cm (1in) below the rim of the container. Remove the plants from their pots by inverting them and giving the base of the pots a sharp tap. Arrange them in their planting positions.

4 Fill the container with compost, pressing it down between and around both plants until it is the correct depth for the daisies and the ajuga. Test this in the same way as before.

5 Remove the two daisies and the ajuga from their pots and arrange them in their planting positions. Fill around them with compost, pressing it down firmly as you go. Top up with compost to 2.5cm (1in) below the rim of the container. Water thoroughly.

Aftercare
Water when the top of the compost feels dry to the touch. Apply a liquid feed fortnightly. Deadhead the daisies regularly. In spring, remove all the plants from the container. Discard the kale and divide the rest to replant either in containers or in the garden.

paprika pepper

The wild abandon of paprika yarrow makes a fabulous basis for a country-style container. Its soft ferny leaves grow upright, producing unusual peppery blooms early in midsummer which fade into a subtle sepia shade and take on a more trailing appearance.

You will need

| Achillea millefolium
 'Paprika'
| Gaillardia 'Dazzler'
| Geum chilense
 'Mrs Bradshaw'
| Corydalis cheilanthifolia
**Large terracotta
 container**
Crocks or small stones
Compost

Season Summer

When to plant Very late
 spring

Site Sunny

❶ Water all the plants thoroughly and allow them to drain. Cover the drainage holes in the container with crocks and add compost 7.5cm (3in) deep.

❷ Position the achillea and gaillardia in the container then fill with compost, pressing it down firmly around the plants.

❸ Add more compost, then position the remaining plants. Top up with compost to 2.5cm (1in) below the rim of the container. Water thoroughly.

Aftercare
Water well whenever the compost is dry to the touch. Apply a liquid feed once a week.

lamplight

This rich combination of creeping and climbing plants lends an interesting texture to an autumn planting. Delicate Chinese lanterns shine out from the sophisticated coppery hues which include exotic Japanese blood grass. Bringing post-summer colour, the ensemble is planted in a wooden tub that has been painted a cool duck-egg blue in contrast to the hot tones of the plants. The plants will flourish until the first frosts when most of them will die back, leaving a lush covering of copper carpet to hide the compost. However, if you then protect the container from frosts, and water it occasionally, it will spring to life again with the advent of warmer days.

❶ Water all the plants thoroughly and allow them to drain. Place crocks over the drainage holes of the tub then add a 10cm (4in) layer of compost.

❷ Position the physalis to one side of the tub and the grape vine on the opposite side, both with their support sticks still in place.

❸ Bed these in, filling in and around each plant with compost and pressing it down well. Add a little more compost and press down firmly.

❹ Arrange the remaining plants in the container and fill in with compost around each one. Add more compost to cover the plants' rootballs and firm.

❺ Once all the plants are firmly planted, carefully remove the supporting sticks of the physalis and the vine and tie them to the tub support with garden string. Water thoroughly.

Aftercare
Water thoroughly whenever the top of the soil becomes dry to the touch. Deadhead the pansies regularly to promote repeat flowering. When the Chinese lanterns of the physalis (which are actually fruit casings rather than flowers) have skeletonized, remove them. When the physalis foliage dies, cut the plant back to ground level. When the vine drops its leaves, cut it back to tidy it and to provide a framework for next year's growth.

Professional know-how
➤ Perennials grown in containers need to be well fed if they are to produce a lush show the following year. Mixing fertilizer granules in the compost when planting up will give the plants a good start. After the first three months, top up regularly with liquid feed during the growing months.

QUIET STAR

Soft and diminutive, the delightful heather, *Erica tetralix* 'Pink Star', looks pretty spreading out in a small copper basket. Plant this lime-hating heather in ericaceous compost, and be sure to water it very regularly as the tiny container will dry out quickly. It should be hung at eye level in a focal position so its delicate beauty can be fully appreciated.

SOFT FOCUS

The gentlest shades can be combined for the most wonderful, subtle results, but there is also a practical reason for grouping several baskets. The heather needs lime-free compost, while the other plants – *Carex berggrenii*, *Sedum kamtschaticum* 'Variegatum' and *Diascia* 'Salmon Supreme' – do not.

alternative plants

The plantings in this book make creative use of plants that are generally widely available. In case any of them prove difficult to locate, however, here is a list of alternative plants that can be substituted. They are listed immediately beneath the plant they can be used to replace, following the order of the original plantings. Alternatively, any of these plants can be used to create your own colour-themed planting ideas.

PRETTY IN PINK

Strawberries and cream - page 22

➤ *Fragaria* 'Strawberries and Cream'
F. 'Pink Panda', F. 'Red Ruby'
➤ *Fragaria* 'Serenade'
F. 'Pink Panda', F. 'Red Ruby'
➤ *Glechoma* (syn. *Nepeta*) *hederacea* 'Variegata'
Hedera helix 'Adam', 'Ester', 'Golden Gate' or 'Sagittifolia Variegata' (syn. 'Ingelise')

Sugar almond shopper - page 24

➤ *Campanula cochleariifolia*
C. GF 'Wilson', C. portenschlagiana, C. 'Birch Hybrid'
➤ *Primula sieboldii*
P. sinensis, P. × scapeosa
➤ *P. whitei* 'Sherriff's Variety' (syn. P. bhutanica)
P. marginata
➤ *P. edgeworthii*
P. allionii, P. petiolaris

Hearts and flowers - page 27

➤ *Dicentra spectabilis*
D. formosa, D. 'Stuart Boothman'
➤ *Tulipa* 'Palestrina'
T. 'Angélique, T. 'Artist', T. 'China Pink'

Ice cream sundae - page 28

➤ *Diascia barberae* 'Ruby Field'
D. barberae 'Rose Queen' or 'Pink Queen'
➤ *Verbena* 'Pink Parfait'
V. 'Amour Light Pink', V. 'Novalis Rose Pink with Eye', V. 'Tapien Pink'
➤ *Fragaria vesca* 'Semperflorens' (syn. F. alpina)
F. 'Pink Panda', F. 'Strawberries and Cream'
➤ *Fuchsia* 'Swingtime'
F. 'Avocet', F. 'Cascade', F. 'La Campanella'
➤ *Fuchsia* 'Pink Galore'
F. 'Avocet', F. 'Cascade', F. 'La Campanella'
➤ *Fuchsia* 'Come Dancing'
F. 'Avocet', F. 'Cascade', F. 'La Campanella'
➤ *Petunia* 'Falcon Red Morn'
P. Surfinia® 'Pink Mini', P. 'Express Blush Pink', P. F1 'Peppermint Daddy'

opposite: Tall, striking and smart, Gaillardia 'Dazzler' makes an ideal focus in large standing containers.

Pink ladies - page 29

➤ *Calluna vulgaris* 'Annemarie'
C. vulgaris 'County Wicklow', 'Finale' or 'JH Hamilton'

Raspberry sorbet - page 31

➤ *Knautia macedonica*
Cosmos atrosanguineus, Scabiosa atropurpurea 'Cockade Series'
➤ *Diascia* 'Lilac Belle'
D. 'Lilac Mist', *D.* 'Pink Queen'
➤ *Brachyscome* 'Strawberry Mousse'
Verbena 'Tapien Pink', *V.* 'Pink Parfait'
➤ *Brachyscome* 'Pink Mist'
Verbena 'Romance Lavender'
➤ *Rhodanthemum gayanum*
(syn. *Leucanthemum mawii*)
Argyranthemum 'Pink Australian'
➤ *Viola* 'Rosy Morn'
V. 'Pink Panther', *V.* Blackberry Rose', *V.* 'Frosty Rose'
➤ *Viola* 'Tutti-frutti'
V. 'Love Duet', *V.* 'Imperial Silver Princess, *V.* F1 'Turbo', wine bicolour'
➤ *Matthiola* Brompton Series
M. 'Cinderella Rose'
➤ *Nemesia caerulea* 'Joan Wilder'
N. caerulea 'Woodcote' or 'Elliott's Variety'

FRESH YELLOWS

Golden girl - page 34

➤ *Alchemilla mollis*
Filipendula ulmaria 'Aurea'
➤ *Lysimachia nummularia* 'Aurea'
Helichrysum petiolare 'Limelight' or 'Roundabout'
➤ *Bidens ferulifolia*
Monopsis lutea, Themophylla tenuiloba
➤ *Argyranthemum* (syn. *Chrysanthemum*) *frutescens* 'Jamacia Primrose'
Coreopsis verticillata, Anthemis tinctoria 'EC Buxton'

Lemon syllabub - page 36

➤ *Mimulus* Malibu Series, yellow
M. 'Magic Yellow', *Calceolaria* 'John Innes'
➤ *Viola* 'Magnum Cream'
V. 'Chantreyland', *V.* 'Princess Cream'
➤ *Viola* 'Sorbet Lemon Chiffon'
V. 'Sunbeam', *V.* 'Prince John'

A crowd of golden daffodils - page 38

➤ *Narcissus* 'Kingscourt'
N. 'Saint Patrick's Day', *N.* 'Unsurpassable'
➤ *Narcissus* 'Spellbinder'
N. 'Golden Harvest', *N.* 'King Alfred'
➤ *Narcissus* 'Jumblie'
N. 'Dutch Master', *N.* Rembrandt'

Hot and sunny - page 40

➤ *Helianthus annuus* 'Teddy Bear'
H. annuus 'Sunspot', *Tagetes erecta* 'First Lady'
➤ *Viola* 'Sunny Boy'
V. Crown Series, yellow splash, *V.* 'Velour Yellow'

Lemon soufflé - page 42

➤ *Crocus chrysanthus* 'Cream Beauty'
C. chrysanthus 'Romance' or 'Snow Bunting'

Tubular belles - page 44

➤ *Digitalis grandiflora*
D. lutea
➤ *Antirrhinum* 'Liberty White'
A. F1 'Bells White', *A.* 'Coronette White'

MOODY BLUES

Flower tree - page 48

➤ *Viola tricolor*
V. 'Sorbet Mixed', *V.* 'Johnny Jump-up', *V.* 'Princess Purple and Gold'

Purple haze - page 50

➤ *Ajuga reptans* 'Multicolor'
A. reptans 'Burgundy Glow' or 'Variegata'
➤ *Aster novi-belgii* 'Professor Anton Kippenberg'
A. thomsonii 'Nanus', *Stachys macrantha* 'Superba'

Blue lagoon - page 52

➤ *Petunia* Surfinia® 'Blue Vein'
P. 'Mirage Lavender', *P.* 'Madness Plum Crazy'
➤ *Convolvulus sabatius* (syn. *C. mauritanicus*)
C. tricolor 'Blue Ensign'
➤ *Lobelia* 'Kathleen Mallard'
L. 'Regatta Lilac', *L.* 'Riviera Lilac'
➤ *Verbena* 'Aphrodite'
V. 'Novalis Rose Pink with Eye', *V.* 'Tapien Pink'
➤ *Nemesia caerulea* 'Joan Wilder'
N. caerulea 'Woodcote' or 'Elliott's Variety'

➤ *Viola tricolor*
V. 'Sorbet Mixed', *V.* 'Johnny Jump-up', *V.* 'Princess Purple and Gold'
➤ *Viola* Universal Series, ivory/rose blotch
Viola Universal Series, white blotch, *Viola* F1 'Turbo', wine bicolour

Growing up - page 55

➤ *Nepeta racemosa* (syn. *N. mussinii*)
Nepeta x *faassenii*, *Veronica peduncularis* 'Georgia Blue'

Chicken in a bucket - page 56

➤ *Muscari armeniacum*
M. azureum, *M. latifolium*, *Bellevalia paradoxa* (syn. *Muscari paradoxum*)
➤ *Iris reticulata*
I. reticulata 'Cantab', 'Clairette' or 'Harmony'

Pretty and witty - page 59

➤ *Salvia officinalis* 'Purpurascens'
Heuchera micrantha diversifolia 'Palace Purple', *H.* 'Rachel, *H.* 'Pewter Moon'
➤ *Viola tricolor*
V. 'Sorbet Mixed', *V.* 'Johnny Jump-up', *V.* 'Princess Purple and Gold'

HOT & SPICY

Coming up roses - page 62

➤ *Lobelia erinus* 'Cascade Red'
L. erinus 'Rosamund' or 'Rose Fountain'

In the red - page 64

➤ *Verbena* 'Nero'

V. 'Amour Scarlet with Eye', *V.* 'Novalis Scarlet with Eye'

➤ *Petunia* Picotee Series, red

P. 'Scarlet Ice', *P.* 'Rose Ice'

➤ *Viola* x *wittrockiana* 'Floral Dance'

V. Universal Series, red, *V.* Universal Series, rose blotch

➤ *Pelargonium* 'Elsi'

P. 'L'Élegante', *P.* 'Variegated La France'

Flaming whoppers - page 66

➤ *Tulipa* 'Apricot Beauty'

T. 'Prince of Austria', *T.* 'Princess Irene'

➤ *Erysimum* (syn. *Cheiranthus*) 'Orange Bedder'

E. 'Fire King', *E.* 'Prince Orange'

Berry bright - page 69

➤ *Gaultheria procumbens*

Cornus canadensis

Warm cinnamon - page 70

➤ *Rosa* Abraham Darby

R. 'Alpine Sunset', *R.* 'Amber Queen', *R.* 'Catherine Mermet'

➤ *Dahlia coccinea* 'Moonfire'

D. 'Bishop of Llandaff', *D.* 'Grenadier'

➤ *Lilium* 'Orange Pixie'

L.. 'Enchantment', *L..* 'OrangeTriumph'

➤ *Mimulus aurantiacus*

Eccremocarpus scaber, *Tropaeolum peregrinum* (with support)

➤ *Hemerocallis* 'Toyland'

H. 'Stella de Oro', *H. minor*

PUREST WHITE & GREEN

Cascade - page 74

➤ *Narcissus* 'Thalia'

N. 'Dove Wings', *N.* 'Portrush'

White magic - page 77

➤ *Viola sororia* 'Albiflora'

V. 'Princess Cream', *V.* Universal Series, white

Chartreuse - page 79

➤ *Helleborus niger*

H. orientalis, white, *H. cyclophyllus*

➤ *Anemone blanda* 'White Splendour'

A. nemorosa 'Vestal' or 'Wilk's Giant'

Round and round the garden - page 84

➤ x *Cupressocyparis leylandii* 'Castlewellan'

Thuja plicata 'Zebrina'

SEPIA SHADES

Rocky road - page 88

➤ *Mimulus* 'Highland Orange'

M. 'Magic Orange', *M.* 'Calypso Mixed'

➤ *Ajuga reptans* 'Rainbow'

A. reptans 'Burgundy Glow' or 'Variegata'

opposite: Dainty Viola tricolor, the wild 'cousin' from which all pansies were originally bred, offers charm to containers from spring to autumn.

Black and tan - page 90

➤ *Eurphorbia griffithii* 'Fireglow'
E. dulcis 'Chameleon'
➤ *Trifolium repens* 'Purpurascens'
Oxalis triangularis 'Atropurpurea'
➤ *Viola* 'Molly Sanderson'
V. 'Bowles' Black', *V.* 'Prince Henry'
➤ *Viola* Universal Series, orange blotch
V. F1 'Turbo', orange, *V.* 'Spanish Sun', *V.* 'Jolly Joker'
➤ *Papaver rhoeas*
Escholzia californica, E. caespitosa 'Apricot Flambeau'

Copper and candy - page 93

➤ *Heuchera micrantha diversifolia* 'Palace Purple'
H. 'Rachel', *H.* 'Pewter Moon'
➤ *Ajuga reptans* 'Burgundy Glow'
A. reptans 'Rainbow' or 'Variegata'

Paprika pepper - page 94

➤ *Gaillardia* 'Dazzler'
Gazania 'Daybreak Bronze'
➤ *Geum* 'Mrs J Bradshaw'
Potentilla 'Gibson's Scarlet', *P.* 'Fireflame'
➤ *Corydalis cheilanthifolia*
C. nobilis, C. wilsonii

Lamplight - page 96

➤ *Imperata cylindrica* 'Rubra' (syn. 'Red Baron')
Melica altissima 'Atropurpurea', *Pennisetum orientale*
➤ *Vitis vinifera* 'Purpurea'
Vitis coignetiae, Parthenocissus tricuspidata 'Veitchii'
➤ *Acaena microphylla* 'Copper Carpet'
A. microphylla, A. caesiiglauca
➤ *Viola* 'Super Chalon Giants', rust
V. 'Imperial Antique Shades', *V.* 'Romeo and Juliet', mixed

Quiet star - page 98

➤ *Erica tetralix* 'Pink Star'
E. tetralix 'Hookstone Pink' or 'Rosea'

Soft focus - page 99

➤ *Diascia* 'Salmon Supreme'
D. 'Ruby Field', *D. barberae* 'Blackthorn Apricot'

index

index

ACKNOWLEDGEMENTS

Debbie Patterson and Tessa Evelegh would like to thank the
following for their assistance:

Gwyn Perry, head gardener at Heale House, Wiltshire, for
checking the text for horticultural accuracy and compiling the list
of colour-coordinated plant alternatives on pages 101-106.

Angie Hicks for planting up and looking after some of the
containers. Ivan Hicks for letting us photograph at Garden in the
Mind at Stanstead Park, Rowlands Castle, Sussex.

Drusilla Stewart of All Seasons Garden and Landscaping, for
planting up and looking after some of the baskets.

Mary Evans for directing the photography and for all her support.

Jackie Matthews for all her support and for her painstaking editing
of the text.

Suppliers

Capital Garden Products
Gibbs Reed Barn, Pashley Road, Ticehurst, East Sussex TN5 7HE
Tel: 01580 201092
Antique bronze-look fibreglass containers

Clifton
Little Venice, 3 Warwick Place, London W9 2PS
Tel: 0171 289 7894
Wirework baskets, antique terracotta pots

Rayment
The Forge, Minster, Thanet, Kent CT12 4HE
Tel: 01843 821628.
Victorian-style wirework hanging baskets

Caroline Pakenham
The Green Man, The Old Manor, Rudge, Frome,
Somerset BA11 2QG
Tel: 01373 830 312 Fax: 01373 830 843
Herb hanging baskets